Animals of the Savanna

Joanne Mattern

The Rosen Publishing Group's

READING ROOM
Collection™

New York

Published in 2003 by The Rosen Publishing Group, Inc.
29 East 21st Street, New York, NY 10010

Copyright © 2003 by The Rosen Publishing Group, Inc.

First Library Edition 2003

Book Design: Ron A. Churley

Photo Credits: Cover, p. 1 © M. Colbeck/Animals Animals; pp. 4–5, 6–7, 12–13, 16–17 © Super Stock; pp. 8–9, 20 © Anup Shah/Animals Animals; pp. 10–11 © OSF Partridge/Animals Animals; pp. 14–15 © Robert Winslow/Animals Animals; p. 18 © Steve Kaufman/Peter Arnold, Inc.; p. 19 © Anthony Bannister/Animals Animals; p. 21 © Phyllis Greenberg/Animals Animals.

Library of Congress Cataloging-in-Publication Data

Mattern, Joanne, 1963-
  Animals of the savanna / Joanne Mattern.
     p. cm. — (The Rosen Publishing Group's reading room collection)
Summary: Introduces the African savanna and some of the wildlife that can be found there, including giraffes, lions, baboons, ostriches, and elephants.
     ISBN 978-1-4358-8996-5
  1. Savanna animals—Juvenile literature. [1. Savanna animals.] I.
Title. II. Series.
  QL115.3 .M28 2002
  591.748—dc21
                              2001007172

Manufactured in the United States of America

For More Information
Animals of the Savanna
http://www.ahsd25.k12.il.us/Curriculum%20info/africa/svanimals.htm

African Animals You Might See on Safari
http://www.ctap3.org/_lperry/africa/animals.htm

Contents

What Is a Savanna?   4

Zebras   6

Giraffes   8

Other Grazing Animals   11

Lions   12

Cheetahs   14

Giants of the Savanna   16

Baboons   18

Birds of the Savanna   20

Animals in Danger   22

Glossary   23

Index   24

What Is a Savanna?

A savanna is a large area covered with dry grass and some trees. Savannas are found in the middle of Africa. There are two seasons in the savanna: A rainy season and a dry season. The temperature in the savanna is usually between 75 and 95 degrees.

The grass on a savanna can grow more than six feet high. Grass is the main food for **herds** of animals such as zebras, buffalo, **wildebeests** (WILL-duh-beests), and **gazelles**. Meat eaters, such as lions and cheetahs, hunt the larger **grazing** animals.

Zebras and wildebeests graze on the dry grass of the savanna.

Zebras

Zebras usually live together in family groups or herds. A male zebra is the leader of each herd. A herd can be as small as five zebras or as big as several hundred. Herds of zebras often graze on the savanna grass with giraffes, wildebeests, and elephants.

Each zebra has a different stripe pattern on its body.

A zebra's stripes help the zebra to hide from enemies that hunt in the tall grass. When zebras crowd together, their enemies can't tell where one zebra ends and another begins. Zebras can run as fast as forty miles an hour to get away from enemies like the lion.

Giraffes

Giraffes are the tallest animals on Earth. Baby giraffes are about six feet tall and can walk about an hour after they are born. Giraffes can grow to be as tall as eighteen feet. Finding food is easy because they can reach leaves high in the trees with their long necks. Giraffes can run as fast as thirty-five miles an hour.

Giraffes eat leaves from trees, shrubs, and sometimes vines.

Wildebeest

Wildebeests are one of the most common animals seen on the African savanna.

African buffalo must drink water at least once a day, after grazing on the savanna.

Other Grazing Animals

The African buffalo has been called the most dangerous land animal. The buffalo uses its strong body, heavy hooves, and big horns to protect itself.

Wildebeests travel thousands of miles across the savanna in search of grass to eat. Baby wildebeests can run a few hours after they are born. This helps them keep up with the herd.

Animals of the savanna usually graze in large herds to stay safe. When one animal is in danger, the whole herd will gather around to guard it.

Lions

Lions belong to a family of big cats. They are the only big cats that live in large groups. These groups are called **prides**. A pride of lions can include thirty females, their cubs, and one or more adult males.

Lions spend most of their time sleeping.

They hunt every few days. A group of lions can hunt and kill a much larger animal, such as a zebra. Female lions do most of the hunting. Males guard the pride. A lion can eat up to seventy-five pounds of meat at one time.

Lions rest or sleep as many as twenty hours a day. Female lions hunt at night for food.

Cheetahs

Cheetahs are another kind of big cat. Cheetahs are the fastest animals on land. They can run more than sixty miles an hour, about as fast as a car on the highway. A cheetah's legs are long and powerful. Its long claws help it get a strong grip on the ground.

Cheetahs run so fast that they can reach a speed of forty-five miles an hour in two seconds!

Cheetahs live by themselves or in small groups. During the hottest part of the day, they lie in the grass under a tree to keep cool.

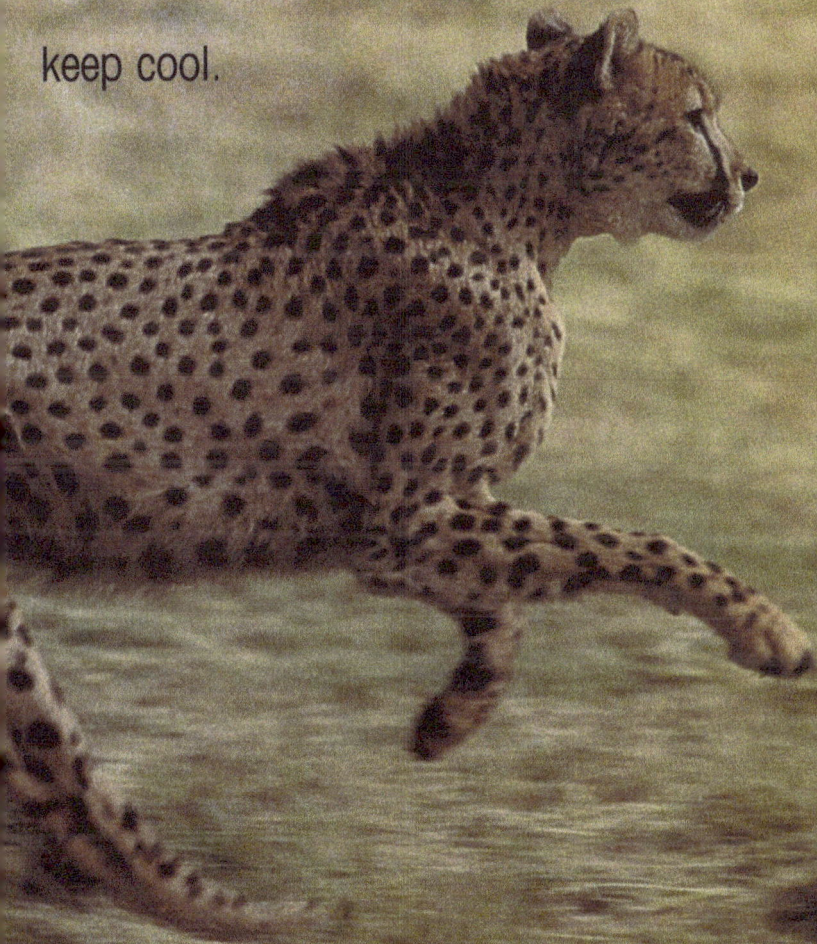

Giants of the Savanna

The African elephant is the largest land animal. It takes a lot of food to feed such a large animal. Elephants spend up to sixteen hours a day eating roots, leaves, bark, and plants. An elephant may eat over 300 pounds of food in a day!

An elephant can live to be over sixty years old and a rhinoceros can live to be fifty years old.

Elephants

The black rhinoceros is another large savanna animal. This animal is about twelve feet long and weighs about 2,300 pounds. The black rhinoceros has an upper lip that is shaped like a hook. It helps the rhinoceros pull leaves and twigs from plants.

Baboons

Baboons are monkeys. They live in large groups called **troops**. Troops travel around the savanna looking for food. A baboon eats many different things, such as insects, worms, eggs, small animals, fruit, and plants.

All members of a troop sleep in the same tree.

Baboons are very smart. They have many different ways of **communicating** with each other. When two baboons see each other, they touch noses as a sign of friendship. They also talk to each other by making many different sounds. Baboons make barking noises when something alarms them and screeching sounds as a warning to other animals they want to scare away.

A troop of baboons can be small or have as many as 200 members.

Birds of the Savanna

Many birds live in the African savanna. The ostrich is the largest bird on Earth. Male ostriches can be nine feet tall and can weigh up to 345 pounds. Ostriches cannot fly because they are so big and heavy, but their long legs help them run very fast.

Ostrich

Male weaver birds use long pieces of grass to weave themselves into a round nest. The nest hangs

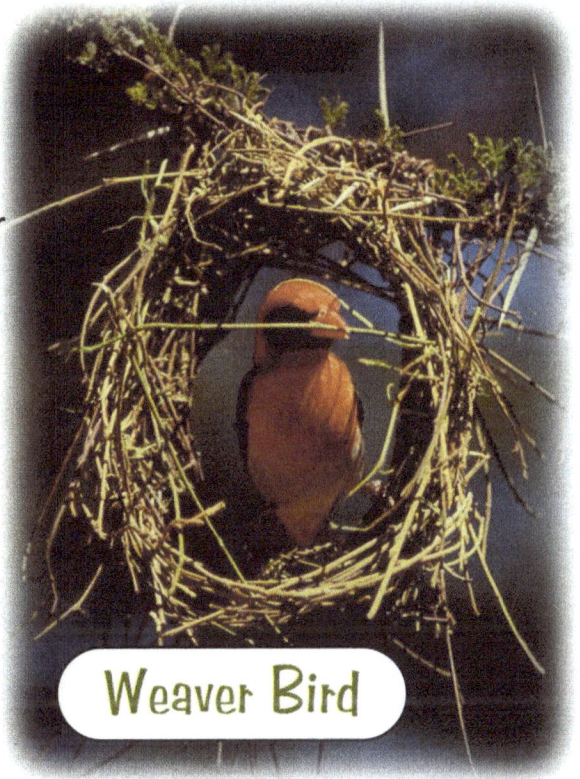

Weaver Bird

at the end of a branch. Some weaver birds build large nests for a community of birds. These nests may have more than one hundred different rooms in them!

The ostrich and the weaver bird are just two of the many birds that live in the African savanna.

Animals in Danger

Savanna animals are in danger. Large areas of the savanna are disappearing. People use the land to build houses, farms, and roads. Animals can no longer live there.

Savanna animals also face danger from **poachers**. Thousands of elephants and rhinoceroses are killed every year for their **tusks** and horns. People use these tusks and horns to make **ivory** jewelry or to make medicine. To protect the animals from poachers, many animals have been moved to **wildlife preserves** where they can live safely.

Glossary

communicate	To share information or feelings.
gazelle	A graceful animal that looks like a deer and lives in Africa and Asia.
graze	To feed on grass.
herd	A group of animals of one kind that graze and travel together.
ivory	A hard white material that makes up the tusks of elephants.
poacher	A person who kills animals illegally.
pride	A group of lions that live together.
troop	A group of monkeys that live together.
tusk	A very long pointed tooth that sticks out of an animal's mouth. Elephants and walruses have tusks.
wildebeest	A large African antelope with a mane, a beard, curved horns, and a long tail.
wildlife preserve	An area where animals are protected from hunters and poachers.

Index

A
Africa(n), 4, 11, 16, 20

B
baboon(s), 18, 19
buffalo, 5, 11

C
cheetah(s), 5, 14, 15

E
elephant(s), 6, 16, 22

G
gazelles, 5
giraffes, 6, 8
grass, 4, 5, 6, 7, 11,
    15, 21

H
herd(s), 5, 6, 11

I
ivory, 22

L
lion(s), 5, 7, 12, 13

O
ostrich(es), 20

P
poachers, 22
pride(s), 12, 13

R
rhinoceros(es), 17, 22

T
troops, 18
tusks, 22

W
weaver birds, 21
wildebeests, 5, 6, 11
wildlife preserves, 22

Z
zebra(s), 5, 6, 7, 13

www.ingramcontent.com/pod-product-compliance
Lightning Source LLC
Chambersburg PA
CBHW041200220326
41597CB00001BA/24